THE INCONTINENT TRUTH
LIES, FRAUD AND BAD SCIENCE

BY ART MAYERS, 2020

IN THE 1950'S CANADIAN OIL-MOGUL MAURICE STRONG CAME TO THE CONCLUSION THAT THE WORLD HUMAN POPULATION WOULD SOON OUTSTRIP GLOBAL NATURAL RESOURCES (NOT A CRAZY IDEA)

AN ESSAY ON THE PRINCIPLE OF POPULATION
THOMAS R. MALTHUS
GREAT MINDS SERIES

TOO MANY PEOPLE, NOT ENOUGH, RAW MATERIALS

ABOUT THE SAME TIME SCRIPPS OCEANOGRAPHER ROGER REVELLE SPECULATED THAT ATMOSPHERIC CARBON DIOXIDE MIGHT CAUSE THE HEATING OF OCEAN TEMPERATURES....

CO2 MAY BE A FACTOR

A YOUNG AL GORE WAS IN ONE OF REVELLE'S CLASSES. SCORED A D-
REPORT CARD
BY THE 1980'S REVELLE SAID CO2 PROBABLY WAS NOT A CAUSE OF TEMPERATURE INCREASES

THEN CAME THE 1970'S
SORRY
NO
GAS

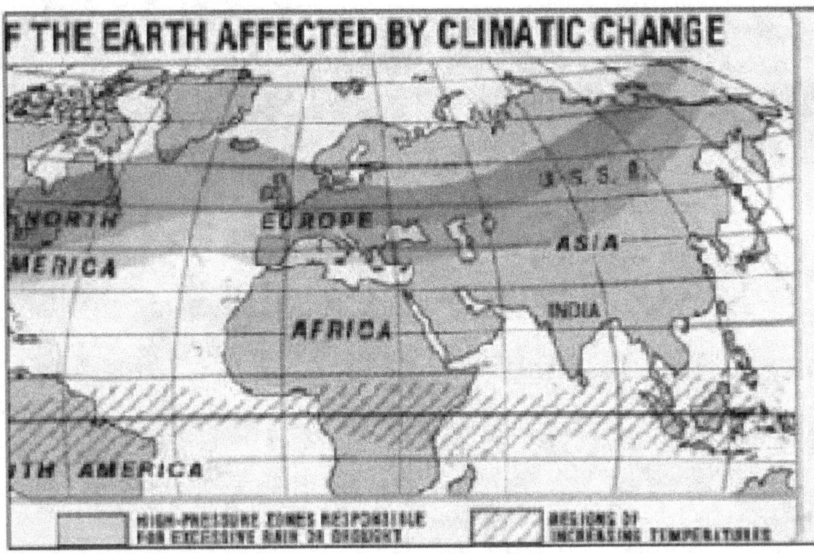
F THE EARTH AFFECTED BY CLIMATIC CHANGE
NORTH
AMERICA
EUROPE
U.S.S.R.
ASIA
AFRICA
INDIA
TH AMERICA
HIGH-PRESSURE ZONES RESPONSIBLE
FOR EXCESSIVE RAIN OR DROUGHT
REGIONS OF
INCREASING TEMPERATURES

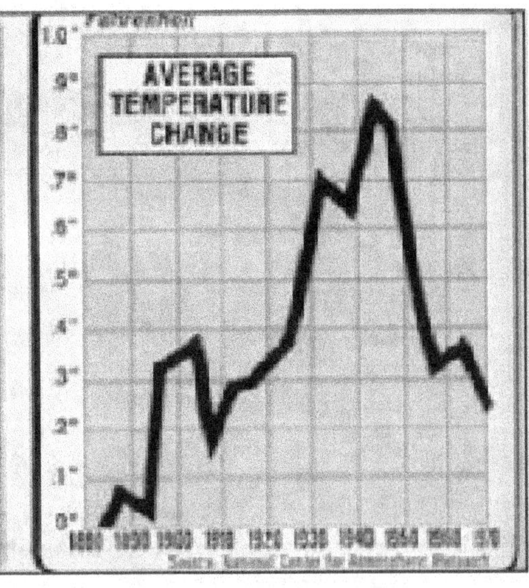
Fahrenheit
AVERAGE
TEMPERATURE
CHANGE
1.0°
9°
8°
7°
6°
5°
4°
3°
2°
1°
0°
1880 1890 1900 1910 1920 1930 1940 1950 1960 1970
Source: National Center for Atmospheric Research

STOP
95.9
95.9

87.9
Self Serve
stock photo

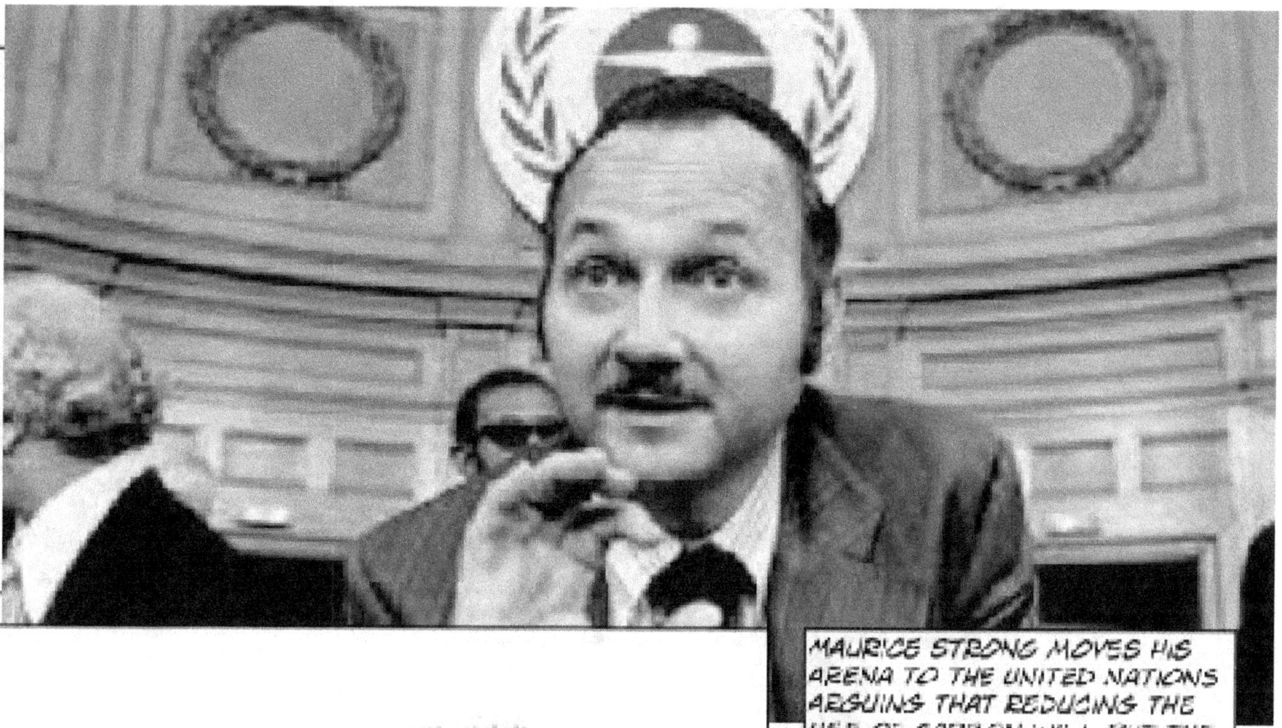

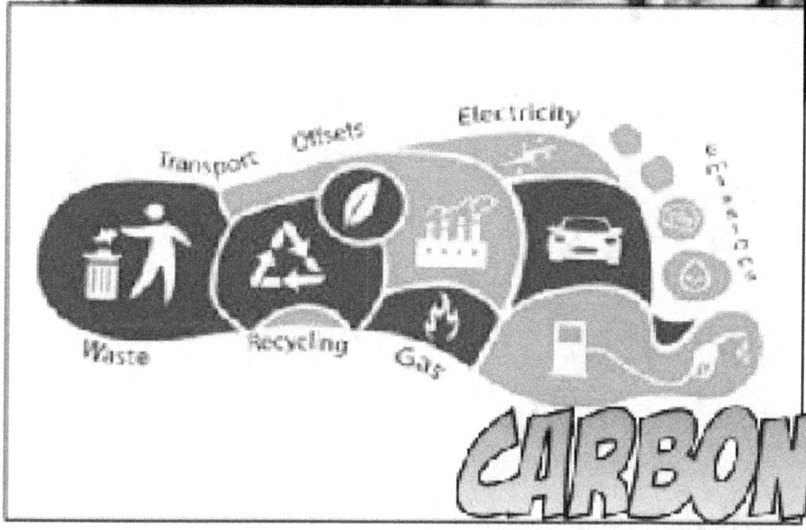

CARBON FOOTPRINT

OZONE HOLE

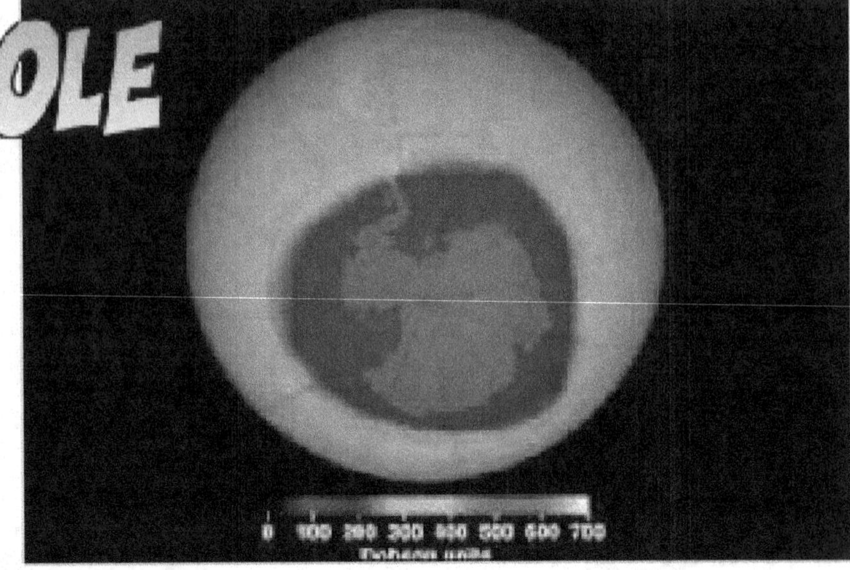

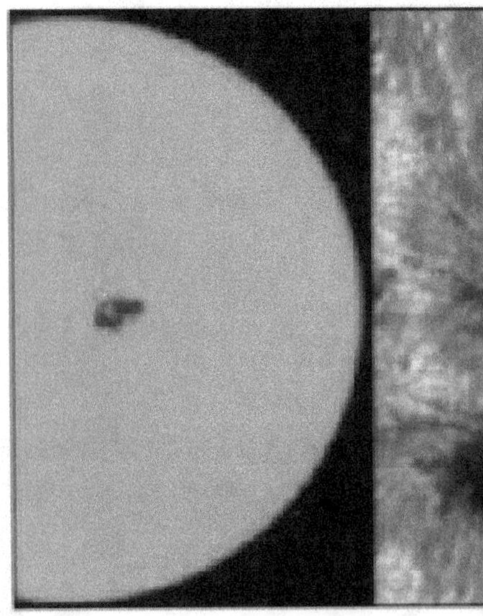

Start of an ozone hole likely in North America

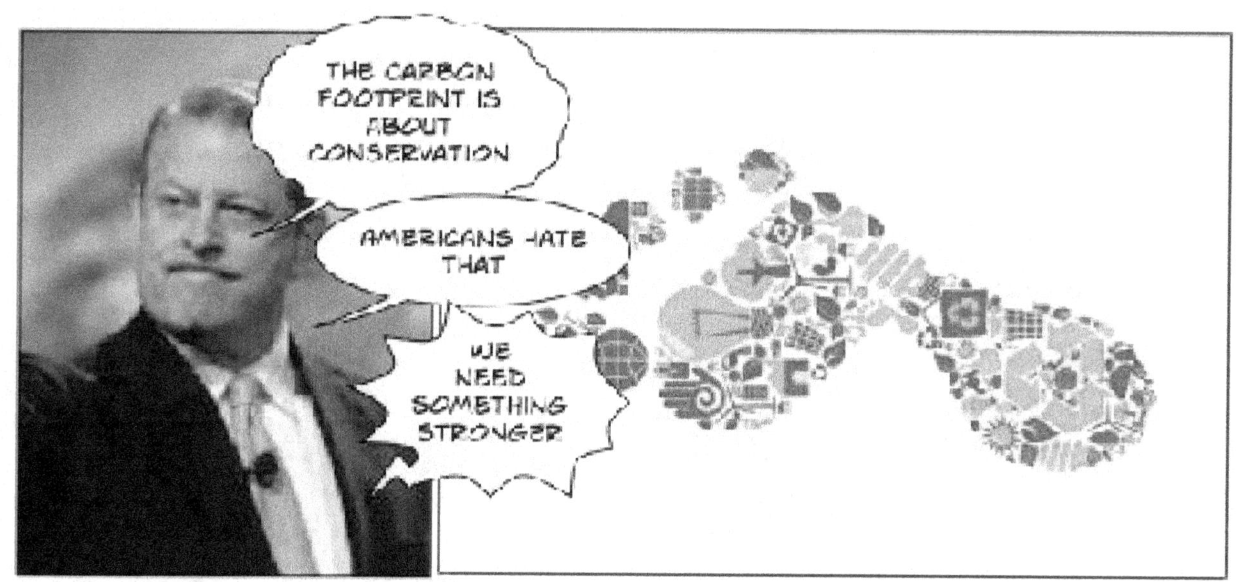

THE CARBON FOOTPRINT IS ABOUT CONSERVATION
AMERICANS HATE THAT
WE NEED SOMETHING STRONGER

WE NEED FEAR

STARVING POLAR BEARS

CITIES UNDERWATER.

ICEBERGS MELTINHG

(Sucrose)
molecule
THEN IN 2006 CARBON, THE BUILDING BLOCK OF LIFE AND THE SOURCE OF ALL STORED ENERGY ON EARTH....
BECAME...
Glucose
Carbon
Oxygen
Hydrogen

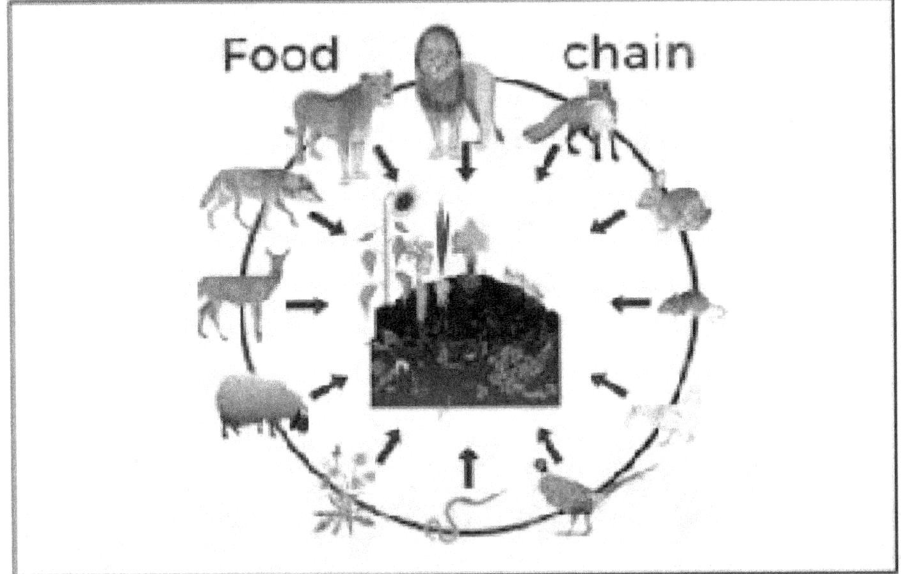

Food chain

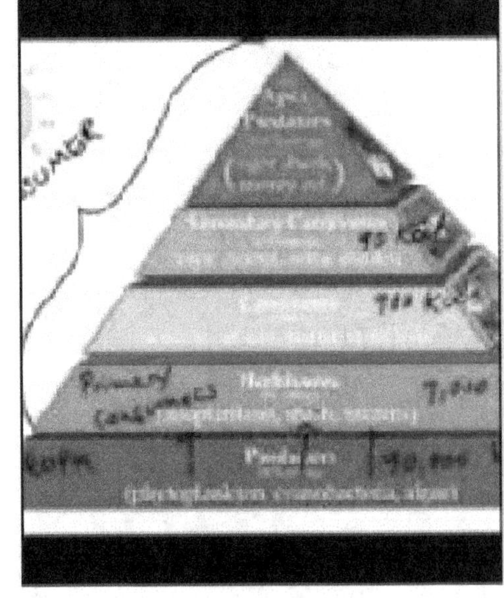

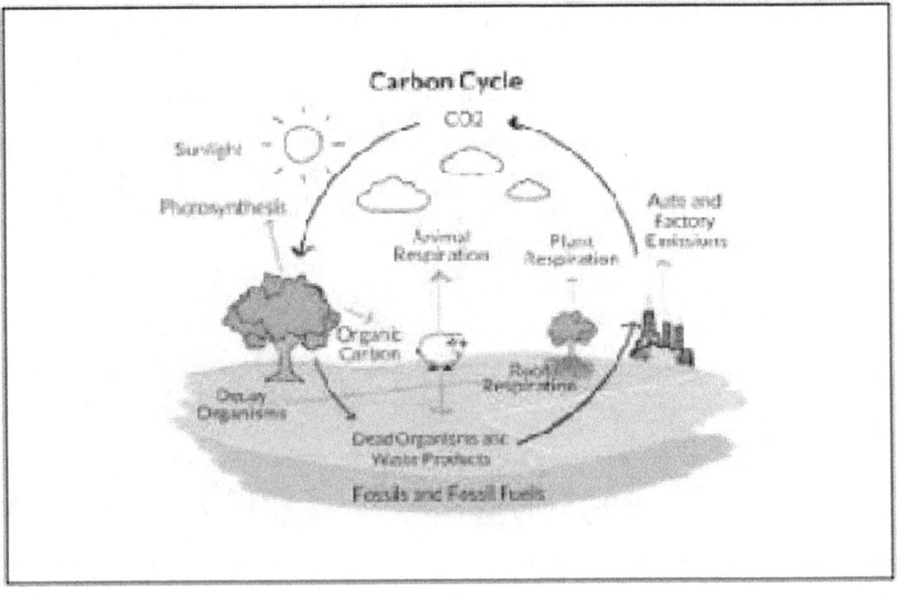

Carbon Cycle
CO2
Sunlight
Photosynthesis
Animal Respiration
Plant Respiration
Auto and Factory Emissions
Organic Carbon
Root Respiration
Decay Organisms
Dead Organisms and Waste Products
Fossils and Fossil Fuels

AHCHOO!
LIKE BLAMING A SNEEZE.........

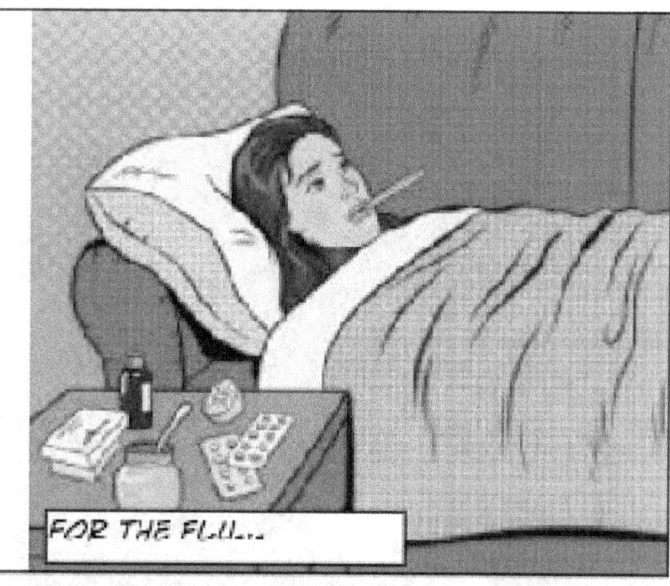

FOR THE FLU...

FIRE ALARM
THE FIRE ALARM FOR THE FIRE......

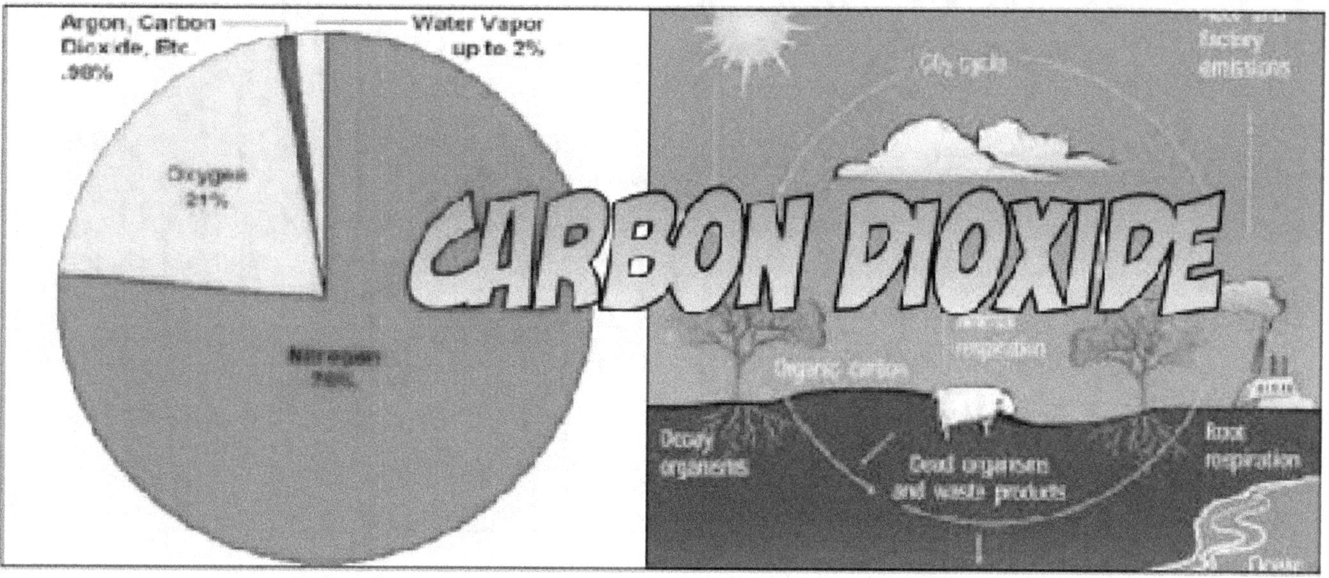

Argon, Carbon Dioxide, Etc. .96%
Water Vapor up to 2%
Oxygen 21%
Nitrogen 78%
CARBON DIOXIDE
Plant and factory emissions
CO₂ cycle
Organic carbon
respiration
Decay organisms
Dead organisms and waste products
Root respiration

AL GORE, FAILED
SCIENCE STUDENT

OIL MAN, MAURICE
STRONG

THE INCONVENIENTRUTH

THE FAILED SCIENCE
STUDENT, GORE, ARGUED
THAT MANMADE CO2 WAS
CREATING EXCESS
GREENHOUSE GASES THAT
WERE INCREASING THE
AVERAGE GLOBAL
ATMOSPHERIC
TEMPERATURES THAT
WOULD LEAD TO A
GLOBAL DISASTER, BUT SO
WAS......

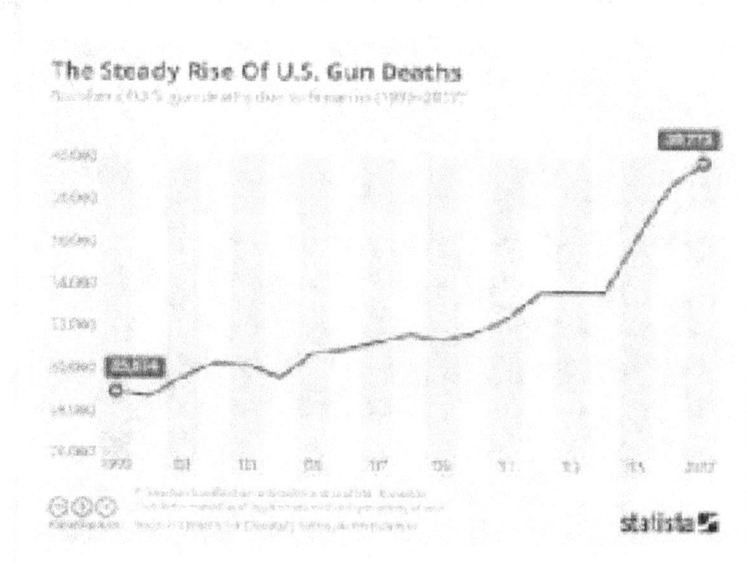

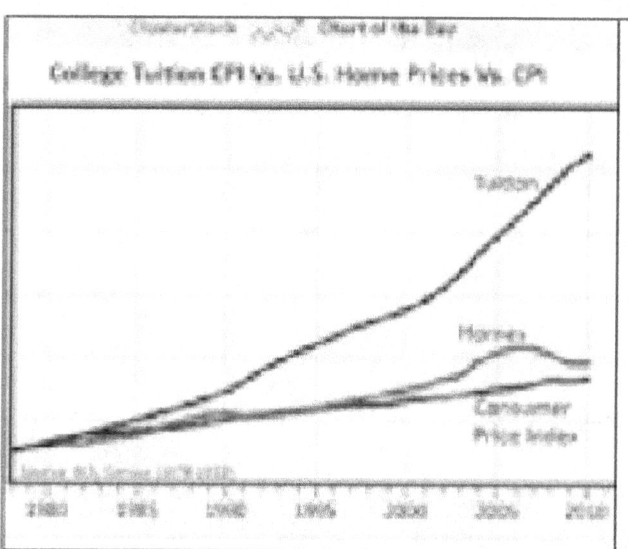

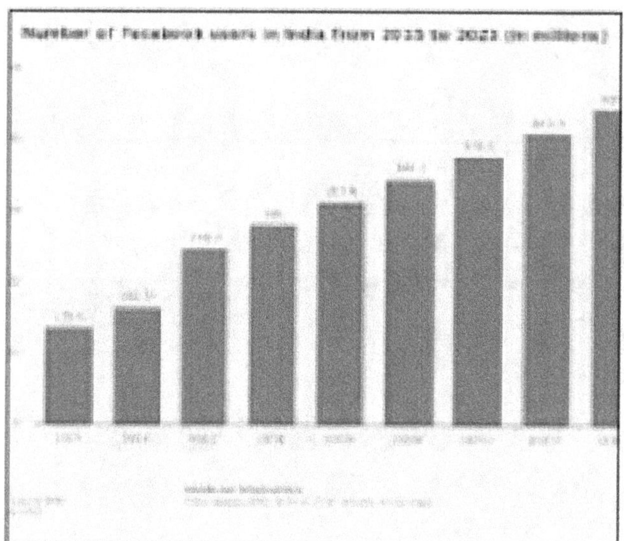

A PHILOSOPHER HAS
STATED THE
CORRELATION BETWEEN
EVENTS DOES NOT
EQUATE TO CAUSATION,
WHY THEN DO 97% OF
SCIENTISTS AGREE...

WHEN DID MY IPAD BECOME A COMPUTER SCREEN?
WHEN DOES MY GRANT MONEY GET HERE?
LOOK AT THAT POLAR BEAR EAT A SEAL
WHAT WAS THE QUESTION AGAIN?
AL WHO?
I'M NOT MOVING OVER

ONCE DOOM WAS ALL BUT CERTAIN AND THE UGLY CARBON FOOTPRINT WAS IN THE PAST PEOPLE BEGAN TO RELAX AND ENJOY THE COMING OF DOOOMSDAY. THEY ALSO INVENTED NEW WAYS OF MAKING THINGS WORSE

AND IGNORED REAL POLLUTION..........

PLASTIC BAGS WERE
BANNED......

I am on
PLASTIC
BAG DIET

MORE PLASTIC CHAIRS
WERE SOLD

PEOPLE BOUGHT
LARGER TRUCKS TO
BRING THEIR RECYCLING
TO THE DUMP

BURN, BABY BURN...

2007 US Energy Consumption
DEMOCRATS SAY THEY WOULD REPLACE FOSSIL FUELS BY 2050
I WONT BE PRESIDENT THEN...
WE ARE SOLID ON 2050
GREEN NEW DEAL
BACK THE DEAL
I MAY BE DEAD BY THEN
I GUESS I WILL BE DEAD

MEANWHILE, STRONG AND GORE
MADE A FORTUNE TRADING
CREDITS FROM POLLUTERS FOR
DUBIOUS CARBON REDUCING
PROJECTS SUCH AS RAIN FOREST
PRESERVATION......

WE TAKE THE MONEY AND MOVE TO THE NEXT LOT

MY CARBON WILL BE HERE FOR HUNDREDS OF YEARS

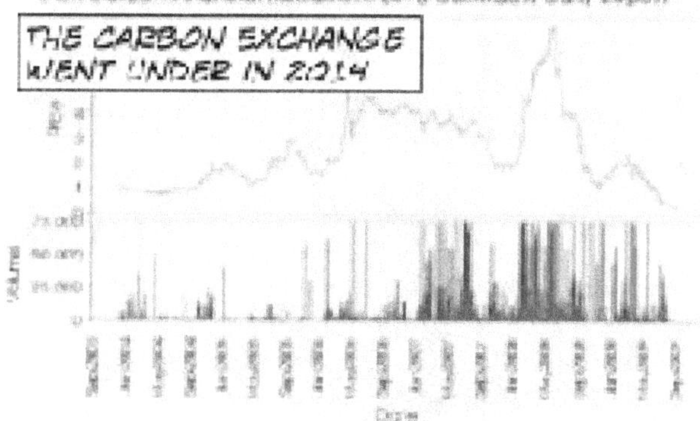

CCX Carbon Financial Instrument (CFI) Contracts Daily Report
THE CARBON EXCHANGE WENT UNDER IN 2014

GORE MADE MONEY AS A SPEAKER.....

STRONG MOVED TO CHINA.........

TO HELP DEVELOPMENT THERE

MEANWHILE...

AS THE DEBATE GOES ON BETWEEN MEATLESS BURGERS, THE NUMBER OF CAMERAS NEEDED ON A CELL PHONE AND WHETHER IT IS RUDE TO TEXT ON A FIRST DATE...

ALTERNATIVE WHAT??

ELECTRIC CAR

GAS FIRED POWER PLANT

HYDROGEN FUEL CELL

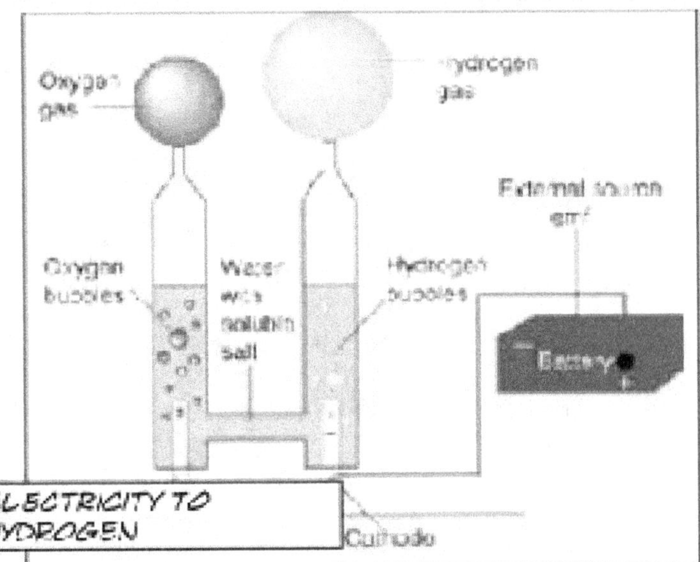

Oxygen gas
Hydrogen gas
External source emf
Oxygen bubbles
Water with soluble salt
Hydrogen bubbles
Battery
Cathode
ELECTRICITY TO HYDROGEN

OOPS

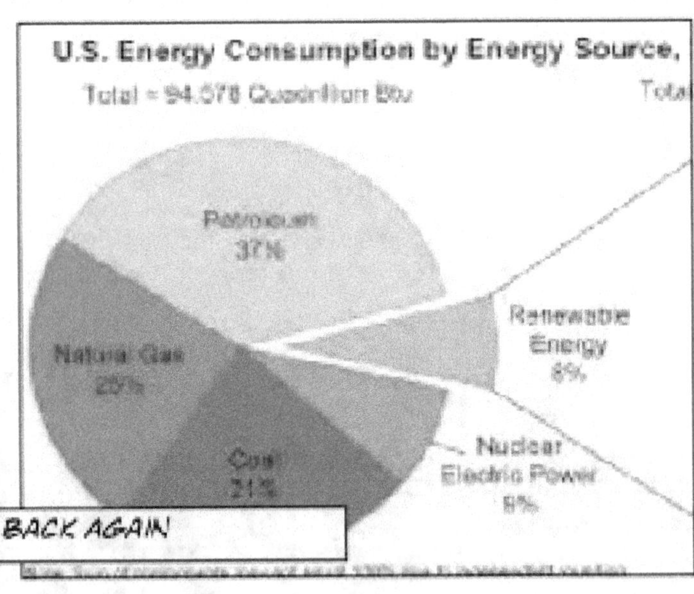

U.S. Energy Consumption by Energy Source,
Total = 94.078 Quadrillion Btu
Total
Petroleum
37%
Natural Gas
26%
Coal
21%
Renewable Energy
8%
Nuclear Electric Power
9%
BACK AGAIN

fossil fuel consumption
WE CONSUME MORE

COSTCO

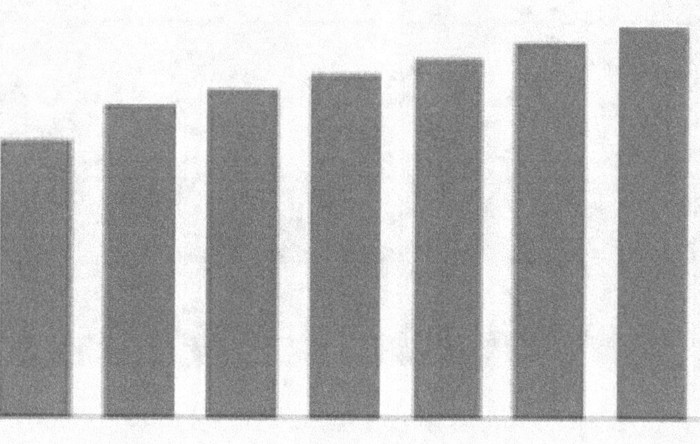

FIRES, EXPLOSIONS

WARS FOR OIL..LIBYA

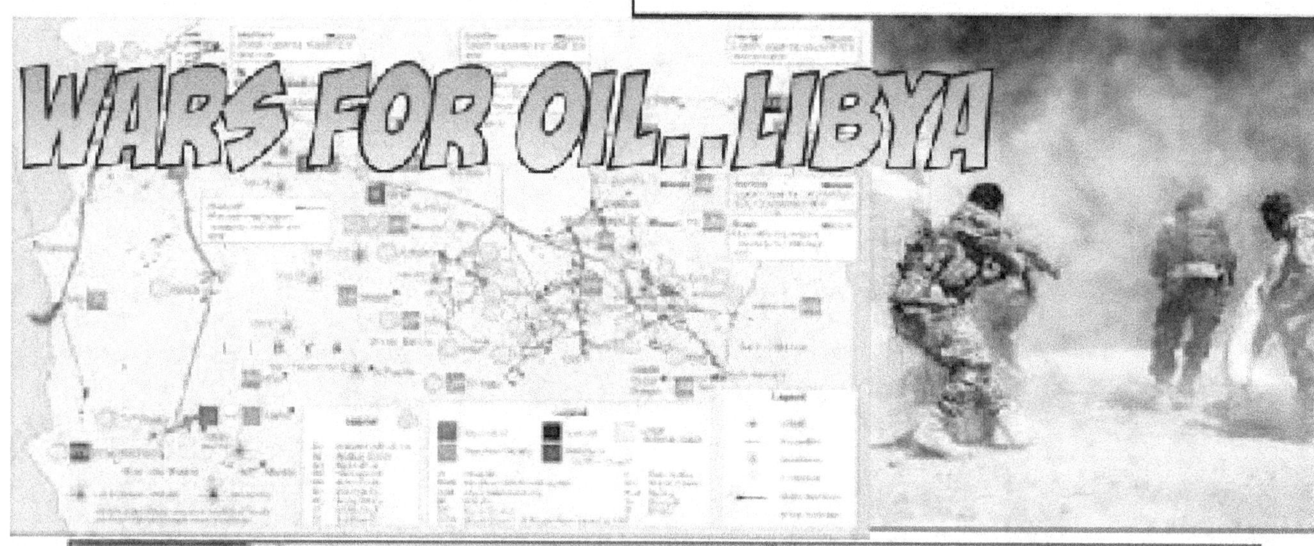

CIVIL UNREST

IRAQ

NOT
ENDLESS
WAR

IRAN

VENEZUELA

GAS THIEVERY
NIGERIA

MEXICO

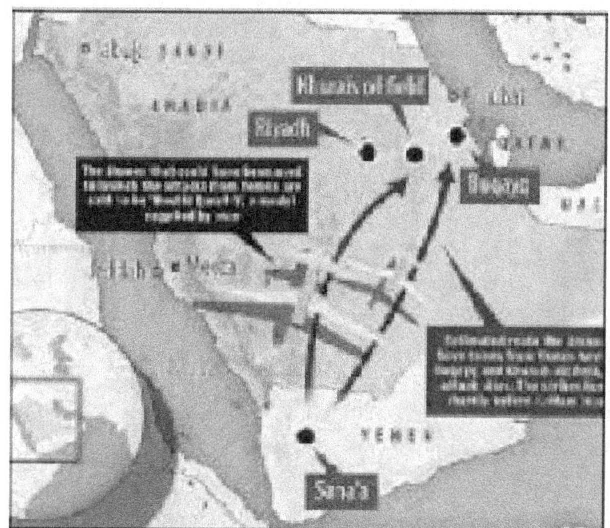

ARABIA
Riyadh
Khurais oil field
YEMEN
Sana'a
QATAR

WAR ON OIL

DANGER
DEEP WATER
KEEP OUT !
POLLUTE MORE

FRACKING

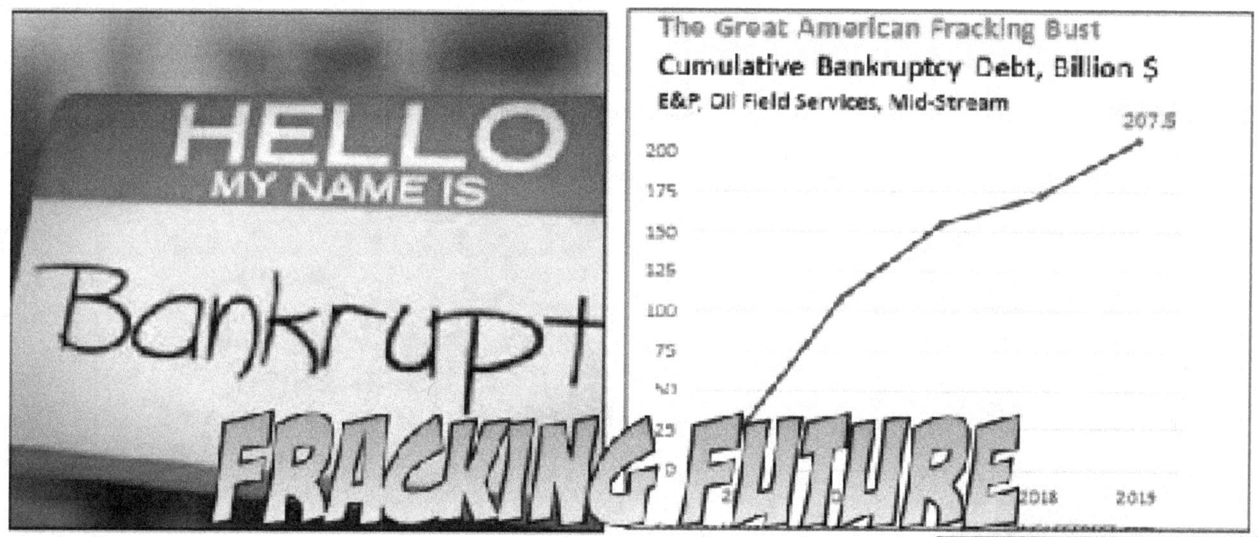

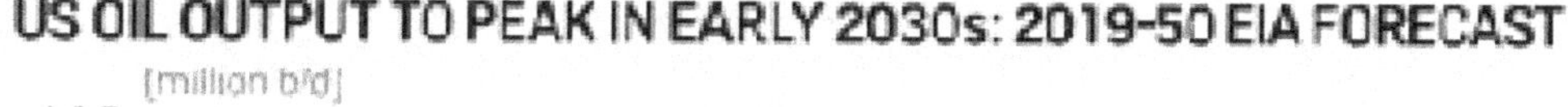

US OIL OUTPUT TO PEAK IN EARLY 2030s: 2019-50 EIA FORECAST

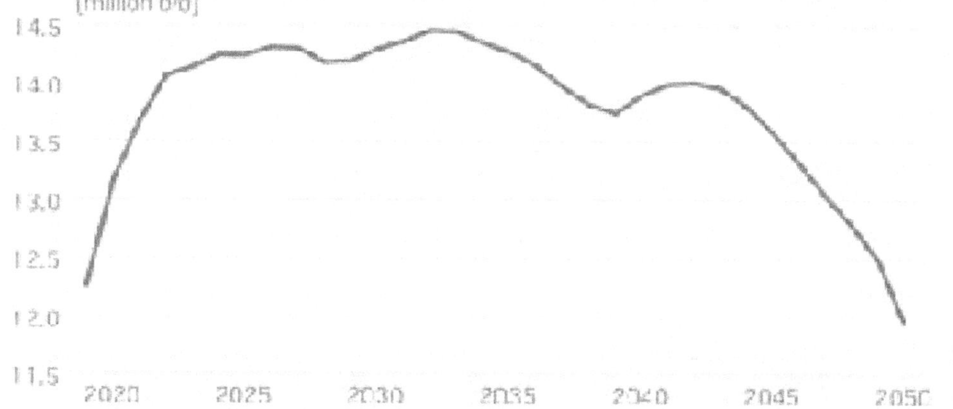

Source: US Energy Information Administration's Annual Energy Outlook

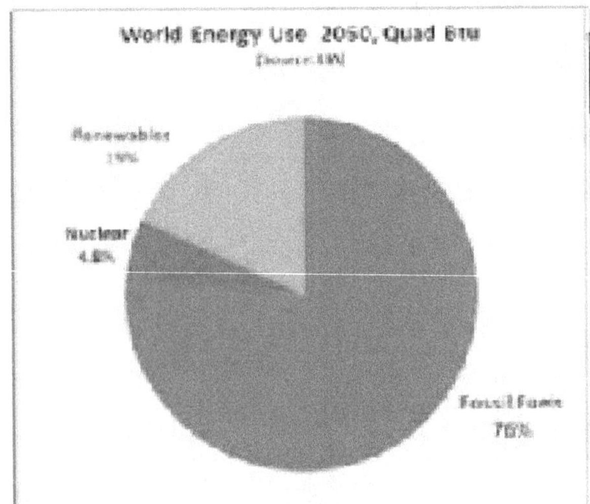

World Energy Use 2050, Quad Btu
(Source: UN)
Renewables
19%
Nuclear
4.8%
Fossil Fuels
76%

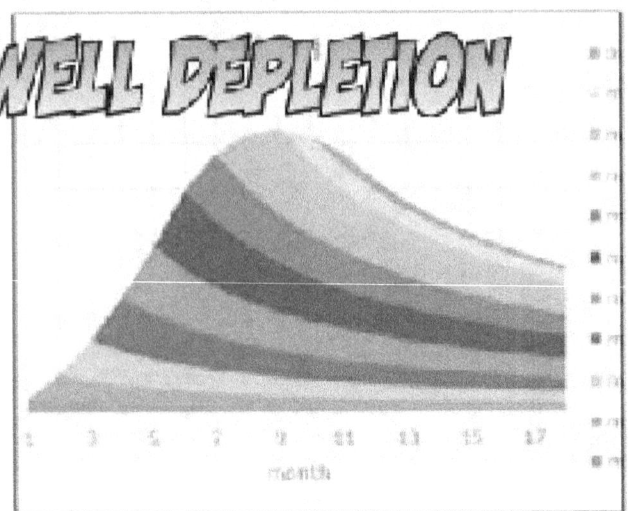

WELL DEPLETION
month

MORE DEBT
LESS ENERGY

NATIONAL DEBT
balance

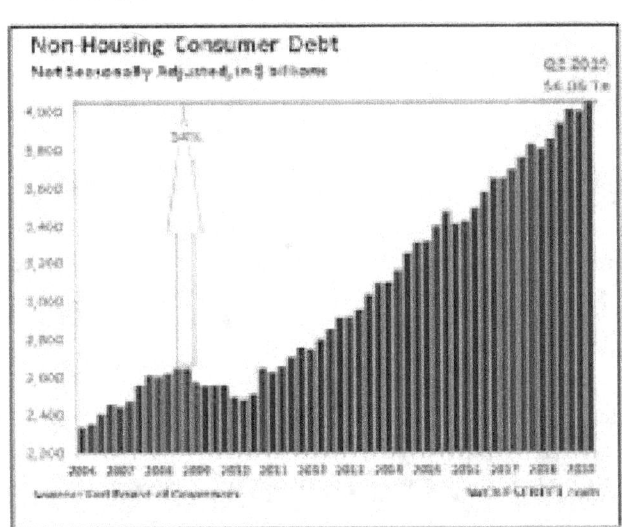

Non-Housing Consumer Debt
Net Seasonally Adjusted, in $ billions
Q2 2020
$4.06 Tn
4,000
3,800
3,600
2,400
3,200
3,000
2,800
2,600
2,400
2,200
2006 2007 2008 2009 2010 2011 2012 2013 2014 2015 2016 2017 2018 2019
Source: Federal Reserve, all Consumers
WOLFSTREET.com

GLOBAL CO₂ LEVELS
CO2 INCREASES....
UAH Satellite-Based
Temperature of the
Global Lower Atmosphere
(Version 6.0)
Running, centered
13-month average
TEMPERATURE SINCE
2014 PRETTY STEADY
WE ARE NOT
SAYING GLOBAL
WARMING
ANYMORE
CLIMATE
CHANGE WORKS
BETTER
CLIMATE
CHANGE HAPPENS
ANYWAY

I
HOPE WE GET
THE GRANT
RENEWAL

WE SHOULD
CHANGE THE
GRAPHICS

UN
MAURICE STRONG SAW
THE PROBLEM, BUT IT
WAS NOT CO2

CURRENT GLOBAL POPULATION
7.7 BILLION
Global population is projected to hit 8 billion around the year
2023, depending on fertility rates over the next few years.
Click the button below to see a live population counter

POPULATION
COUNTER

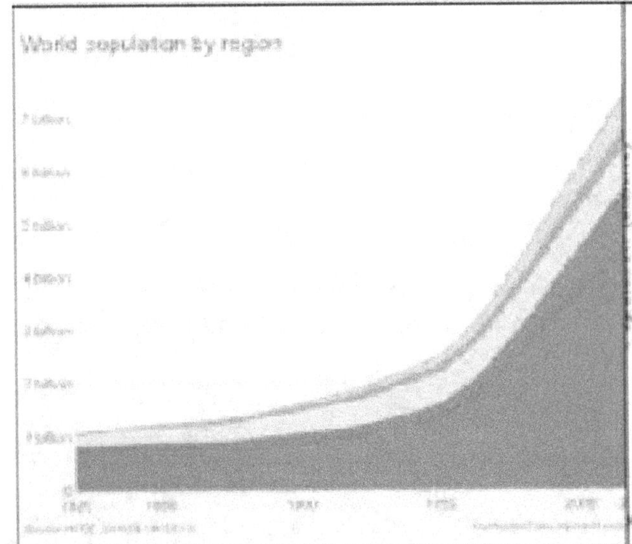
World population by region

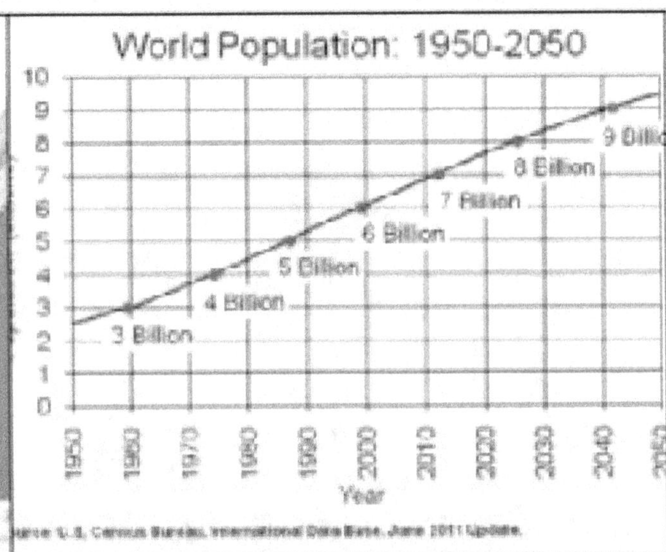
World Population: 1950-2050
9 Billion
8 Billion
7 Billion
6 Billion
5 Billion
4 Billion
3 Billion
Year
Source: U.S. Census Bureau, International Data Base, June 2011 Update.

Without Google Images and Comic Life this book would have taken years to illustrate.

ARCHIMEDES.....
(PARAPHRASE)
WHEN ICE MELTS
IN A GLASS OF
WATER THEN THE
WATER LEVEL
DOES NOT
INCREASE

"IT IS COMMON ERROR TO INFER
THAT THINGS WHICH ARE
CONSECUTIVE IN ORDER OF TIME
HAVE NECESSARILY THE
RELATION OF CAUSE AND EFFECT.

JACOB BIGELOW

OIL IS TOO PRECIOUS A
COMMODITY TO BURN

THE SHA OF IRAN

Art Mayers 2/2020